On Rotation

Greg Feild

August 18, 2018

On the improvement of the understanding:

> . . . ideas which are clear and distinct can never be false : for ideas of
> things clearly and distinctly conceived are either very simple themselves,
> or are compounded from very simple ideas --- that is, deduced therefrom.
> The impossibility of a very simple idea being false is evident to anyone
> who understands the nature of truth or understanding and of falsehood.

⇔

> . . . we may never, while we are concerned with inquiries into actual things,
> draw any conclusions from abstractions; we shall be extremely careful not
> to confound that which is only in the understanding with that which is in
> the thing itself.

⇔

> . . . words are formed according to popular fancy and intelligence,
> and are, therefore, signs of things as existing in the imagination,
> not as existing in the understanding.

-- Benedict de Spinoza

⇐ ⇒ ⇐ ⇒

The discovery of truth is prevented most effectively, not by the false appearance
things present and which mislead into error, nor directly by weakness of the reasoning
powers, but by preconceived opinion, by prejudice, which as a pseudo *a priori* stands
in the path of truth and is then like a contrary wind driving a ship away from land, so that
sail and rudder labour in vain.

-- Arthur Schopenhauer

Abstract:

In this paper, we show all phenomena can be understood as the conservation of mass, angular momentum and rotational kinetic energy.

In addition, we derive the equation for the relativistic mass of a particle, using only Newton's second law and and the kinematic variables defined by the particle wave function, without resort to, or reference to, coordinate transformations.

We also provide a small correction to our model of photon propagation, and continue our critical examination of the concept of potential energy.

Finally, we demonstrate the equivalence of Newtonian and Hamiltonian dynamics.

In the universal model, nature acts to minimize the work done during any interaction.

Die Welt ist alles, was der Fall ist.

-- Ludwig Wittgenstein

Jabberwocky:

> 'Twas brillig, and the slithy toves
> Did gyre and gimble in the wabe:
> All mimsy were the borogoves,
> And the mome raths outgrabe.

Humpty Dumpty:

> "When I use a word," Humpty Dumpty said, in rather a scornful tone,
> "it means just what I choose it to mean -- nothing more nor less."
>
> "The question is," said Alice, "whether you *can* make words mean so many things."
>
> "The question is," said Humpty Dumpty, "which is to be master -- that's all."   :)

-- Lewis Carroll
Through the Looking-Glass

One should not wrongly *materialize* "cause and effect," as the natural philosophers do
(and whoever like them naturalize in thinking at present), according to the prevailing
mechanical doltishness which makes the cause press and push until it "effects" its end:
one should use "cause" and "effect" only as pure *conceptions*, that is to say,
as conventional fictions for the purpose of designation and mutual understanding, --
not for explanation.

-- Friedrich Nietzsche
1886

Neologisms:

We've coined a lot of phrases during the development of "the universal model", mostly on the fly and without too much forethought:

- neutrinium
- the universal model
- the total coupling charge; tcc
- quantum mechanical electromagnetic induction: qemf
- the tetrahedron (perhaps the polyhedron, polyhadron?)
- the electromagnetic charge
- the mass charge
- the compton radius of the neutrino
- mass isospin
- spinoring
- corpusculating

"The universal model" is pretty dull (and already taken by someone!) but it is a handy adjective; e.g the universal reference frame, the universal principle of equivalence, etc.

What we christened mass isospin should really be called charge isospin, and weak isospin (which we had retained for historical reasons) should really be called mass isospin, if we are are to be true and parallel to the original isospin model for nucleons. We will make and justify these changes later in this book.

The terms we chose for the methods of propagation for the photon and electron, corpusculating and spinoring, are rather long, boring, and clumsy.

What we *meant* to say, was, photons gyre and electrons gimble.   :)

But, take your pick ...

Inventing words is hard!

Errata:

I am not an accomplished speller, although I do know the difference between principle and principal; at least in principle! Unfortunately, over our last several books, we have written the phrase 'principle moment of inertia', too many times to count! I blame the spell checker, which is now, of course, screaming bloody murder. But, it's dun! Moving on …

We will resurrect the ideas of 'the electromagnetic charge' and 'the mass charge' first introduced in "On Parity and Isospin" (along with the idea of 'mass isospin') and later abandoned for technical reasons. These ideas were not working out as we had the concepts of mass and charge isospin exactly backwards, along with the identity of the corresponding conserved currents. Also, of course, our model of the particle magnetic moment was not fully realized at that time.

We also offer a small, but important, clarification on the nature of the photon and how it rolls.

Finally, in all our expressions and formulations we will bring all suppressed variables (i.e. c=1, h^{bar}=1) to the fore. To date, we've been carelessly mixing Units, Natural Units, etc.; obviously not without some unnecessary confusion!

Introduction:

Relativistic quantum mechanics may be considered the theory of everything.

Rotation, in all its guises; as translation, angular momentum, spin, helicity, and precession, is the basis for all particles and all particle interactions.

Angular momentum is *the* fundamental quantity, and quality of matter, and the only constant of nature.

Angular momentum is *more* fundamental than even mass or electric charge, since these quantities are essentially emergent properties of particle spin.

The fundamental unit of angular momentum is Planck's quantum of action; h.

Everything reduces to spin. Spin is absolute motion. Spin is contagious.

Spin drives the universe.   Spin is in.

The most fundamental, irreducible quantity we can assign to an elementary particle is spin. A particle is essentially spin at a point in space and time.

One cannot deny spin!

gf

:)

Newton's laws:

Our previous reflections on Newton's laws and the Lorentz force are provided in Appendix A for ease of reference.

In Appendix A, we generalized, or univeralized, Newton's third law from $\mathbf{F}_1 = -\mathbf{F}_2$ to $(action)_1 = -(action)_2$, or mathematically speaking;

$$F_1 \times d_1 \times t = -F_2 \times d_2 \times t \tag{1}$$

Equation (1) states that the *work done* by the particles, each on the other, Is equal and opposite, rather than the forces. The forces are *only* equal in magnitude, if the two objects each move the *same distance* (i.e. $m_1 = m_2$), or for static forces.

So, for example, as depicted in Figure A1, the central force between two bodies in the universal reference frame (i.e. the force on the reduced mass) is not the same as the individual forces of each body on the other, but their sum.

We must conclude Newton's third law ($\mathbf{F}_1 = -\mathbf{F}_2$) is incorrect, even for most *central* forces.

I know ! Right ? . . . we, too, are a little shocked, surprised, and unsettled . . .

However, the actions of the two particles are equal and opposite. Action is always conserved.

In order to fully understand the nature of two body motion and the forces involved, we will recast Newton's three laws of motion (of Appendix A!) solely in terms of angular momentum. We believe angular momentum to be *the* fundamental concept or quantity of physics, even more fundamental than linear momentum, which is only a limiting case (… we always learn it backwards in school!).

Linear momentum is a human-sized, *earthbound* concept, as are the ideas of force, rest, and the notion that objects do not have the propensity to spin. If we did not inhabit human bodies on a flat earth, we should probably believe none of these things!

In addition, and 'as we know', one may always 'transfer linear momentum away'. The conservation of linear momentum attests to the fixity of, and the isotropic, homogeneous, etc. nature of, space and time. There is no more physics there.

Angular momentum is where the action is !   (No pun intended.)
Rotational symmetry is the only symmetry.
Angular momentum is the only conserved quantity and determines all the laws of physics.
                            This is the premise of the universal model.

We will reexpress Newton's laws in terms of angular momentum, rather than linear momentum, with all calculations done in the universal reference frame of Figure A1.

The advantages of this approach are several. When we analyse interactions in the center of mass of the particles of interest, all external torques (arising from conservative gravitational sources) cancel out; these external forces affect only the linear momentum of the center of mass, which is arbitrary.

When we characterize interactions in terms of torques, and the rate of change of angular momentum of each particle in a system of particles relative to an 'arbitrary' point, the results are applicable, and can be generalized, to *all* reference frames, even to *accelerating* reference frames.

That's the power of spin !   :)

Newton's first law becomes:

If a particle does not experience a change in *angular momentum* relative to *any* arbitrarily chosen point in the universal reference frame, then the particle is considered to be free (i.e. there are no net forces acting on it).

For a two body system, if there is no change in the angular momentum of *either body* comprising the two body system relative to *any* arbitrary point (this 'excludes' the choice of points lying along the unit vector, **r**), then the particles are *not interacting*.

Newton's second law becomes:

A particle that undergoes a change in angular momentum relative to our arbitrarily chosen point (i.e. the origin of our coordinate system) is said to experience a net torque, τ ;

$$\tau = dL/dt = \mathbf{r} \times \mathbf{F} \tag{2}$$

where the force, **F**, is defined by equation A1.

For two body 'central force' motion, the torques experienced by each individual body relative to our chosen reference point of Figure A1, are equal and opposite;

$$\mathbf{r_1} \times \mathbf{F_1} = -\mathbf{r_2} \times \mathbf{F_2} \quad ; \quad |\mathbf{F}| = |\mathbf{F_1} - \mathbf{F_2}| \tag{3}$$

Now that we are expressing the forces between interacting bodies in terms of torque, we can include the 'spin-spin' interaction of our theory in a rigorous way into our universal force equation. Previously, we had to *resort to* the concept of potential energy to include this interaction.   : (

We remind the reader that the "Lorentz torque" is defined as the interaction of the moment of inertia of one body with the "gravitational magnetic vector", **B_g**, of the other body, and vice versa.

$$\tau_{SPIN} = I_1 \times B_2 + I_2 \times B_1 \qquad (4)$$

and

$$\tau_{TOTAL} = \tau_{FORCE} + \tau_{SPIN} \qquad (5)$$

Newton's third law becomes:

During a two body interaction, the two bodies will undergo equal and opposite changes in their respective '*actions*'; i.e. they will have equal, and 'opposite', changes in kinetic energy.

$$\int F_1 \cdot dr_1 dt = -\int F_2 \cdot dr_2 dt \qquad (6)$$

Or, since the time is common to both integrals, and the limits of integration are arbitrary;

$$\Delta T_1 = -\Delta T_2 \qquad (7)$$

Newton's universal law of gravitation, expressed in the center of mass of a two body system, becomes (equation A9);

$$F/E_{TOT} = K^*(c/R)^2 \mu - K^*(\mu v^2/R^2) - K^*(l^2/\mu R^3) \qquad (8)$$

where the second term on the right hand side is the *coriolis* force; our answer to spacetime disturbances.

We note, just for fun, that all test particles follow the exact same trajectory in a Newtonian gravitational potential.

However, the concept of potential energy is *so* last-century.

Potential energy:

The concept of potential energy, and that of the force field and the 'test charge', has always been problematic. Potential energy is defined as the work done to move an object a prescribed distance in a conservative force field. How is this work being done and by whom? More importantly, how is this potential energy recognized, *physically acquired*, stored and released by a particle or a system of interacting particles?

It's not clear.

Potential energy is a useful concept for studying closed, conservative systems, like planetary orbits, the hydrogen atom, etc. However, it is *incorrect* to imagine that some nebulous, ethereal, spooky potential energy is in any way "stored" in these bound systems. The energy of these systems are completely determined by summing the individual kinetic energies of the particles Involved.

For example, in an atomic explosion, one is not releasing stored potential energy. One is freeing kinetic energy. The orbital kinetic energy of the bound particles becomes linear kinetic energy once the particles are released.

The concept of potential energy is particularly ill-defined for understanding how these closed, conservative systems form in the first place. Imagine an electron and a proton separated to a distance, $R = \infty - 1$. Is the potential energy between the two particles zero or infinite? Conventionally, we would say zero, however, when we 'let the two particles go', they will start moving toward each other, acquiring ever increasing, equal and 'opposite' kinetic energy on the way.

In the formation of the hydrogen atom, -13.6 eV of work is not done in bringing an electron in, to orbit around a waiting, stationary proton; rather an electron and proton with the correct relative kinetic energies (and momenta, for direction!) will meet and form a bound system. We would not even venture to say that the proton "captures" a passing electron (except as a simplifying approximation for models, etc.), as this would imply a fixed center of force, or "cause", which we *do not allow* in our model of action and reaction.

Particles act on each other, equally, always.

In order to remove to concept of potential energy from our model, but keep the concept of the conservation of energy, we must allow for 'negative' kinetic energy. The kinetic energy would take the sign of the particle momentum.

Kinetic energy is thought to always be positive because it depends on the square of the velocity, however, this is *only* a convention.

Work can be negative. Potential energy can be negative. Now kinetic energy is negative!

Getting rid of potential energy, formally, at least, has several important consequences for putting the photon on equal footing with the leptons. Photons have kinetic energy, but they cannot have potential energy by the usual definition of bringing a particle in from infinity and calculating the work. However, the photon is allowed to execute closed orbits in a potential well, although the photon has to be 'going in the right direction' before capture. Even in an 'infinite potential', a photon cannot reverse linear direction, but rather the photon frequency will tend to zero.

Hamiltonian dynamics:

In our model, the Lagrangian for *any* interaction, classical or quantum mechanical, is the change in kinetic energy of the system;

$$L = \Delta T = \Delta E = \Delta m c^2 \qquad (9)$$

With this definition, when we move to relativistic quantum mechanics, we will no longer need to "kludge" an effective Lagrangian from guesswork, knowing the desired results in advance.

So, our minimization principle is; nature always minimizes the change in kinetic energy integrated over the time of an interaction,

$$\delta \int \Delta T \, dt = 0 \qquad (10)$$

Or, *equivalently*; nature minimizes the work done over the time of the interaction,

$$\delta \int \int_r \mathbf{F} \cdot d\mathbf{r} \, dt = 0 \qquad (11)$$

Besides the obvious lack of an interaction term (!), equation (10) involves scalar quantities, and we are no longer confident about unambiguously assigning such variables a positive or negative value. So, equation (11) is our formal minimization principle.

(We expect this will lead to a minimization principle of least time for light, and least distance for matter, although we've yet to work through the math.)

Since the limits of the time integral in equation (11) are arbitrary, this means nature is *always* minimizing the work, and it would be nice to have an instantaneous (i.e. derivative) expression for our minimization principle.

We want to rid our theory of all things linear (except in the mathematics!), so we will recast equation (11) in terms of the variables of the universal reference frame of Figure A1, except that we will (explicitly) remove the time integral;

$$\delta \int \tau \cdot \mathbf{d\theta} = 0 \qquad (12)$$

where the torque is

$$\tau = dL/dt = \mathbf{rxF} \qquad (13)$$

and if we can write

$$\mathbf{d\theta} = (\mathbf{d\theta}/dt)\, dt = \omega\, dt \qquad (14)$$

then equation (12) becomes

$$\delta \int dL/dt \cdot \omega\, dt = 0 \qquad (15)$$

In equation (15), of course, the integral must be done for object 1 and 2.

Now we have a formulation of Hamilton's principle derived from Newton's laws and identical to that resulting from our new universal Lagrangian, *and* cast in terms of *angular frequency*.

Spoiler alert: Foreshadowing!

The conservation of mass:

The first conservation law was the conservation of mass. Next, was added the conservation of energy. Then, these two concepts were merged to form the conservation of mass-energy.

Now we are back to conservation of mass.

Mass is not converted into energy. It is converted into other mass; either light or matter.

Technically, there is no energy, *only* mass and motion.

Mass is a measure of 'particle angular momentum'.

Consequently, we will always speak of the mass of a photon and 'never' the energy.

There are two kinds of mass; light and matter, or photons and leptons.

All quantum mechanical probability currents express the conservation of mass.

Conservation of mass holds in any reference frame, because mass difference is a scalar and Lorentz invariant.

There is only conservation of mass. We retain kinetic energy for practical purposes.

The fundamental atom:

The fundamental atom, the origin and source of all mass and interaction, is Planck's quantum of action; h.

The photon *is* one quantum of action. However, the photon is *not* the most 'fundamental' particle! The sole source of photons is the acceleration of leptons. Thus, the photon is the quantum of *interaction*.

The sole source of all photons is accelerated matter shedding mass, so to understand the photon and the quantum of action, we really need to understand the nature of the leptons; in particular the spin nature of the neutrino and how it 'acquires' mass. Someday!

But, we start with the photon.

The photon:

The photon is one unit of angular momentum, *the fundamental unit of angular momentum*, Planck's quantum of action; h. The angular momentum of the photon is always h, and the speed is always c. So, how does the photon gain and lose energy and linear momentum during an interaction?

In our model, the projection of the photon angular momentum vector, along the direction of travel, varies sinusoidally at the frequency that defines the energy and momentum of the photon according to the usual relations;

$E = h\upsilon$
$p = h\upsilon/c$ (16)
$m = h\upsilon/c^2$

Let us say the photons 'gyres' as illustrated in Figure 1.

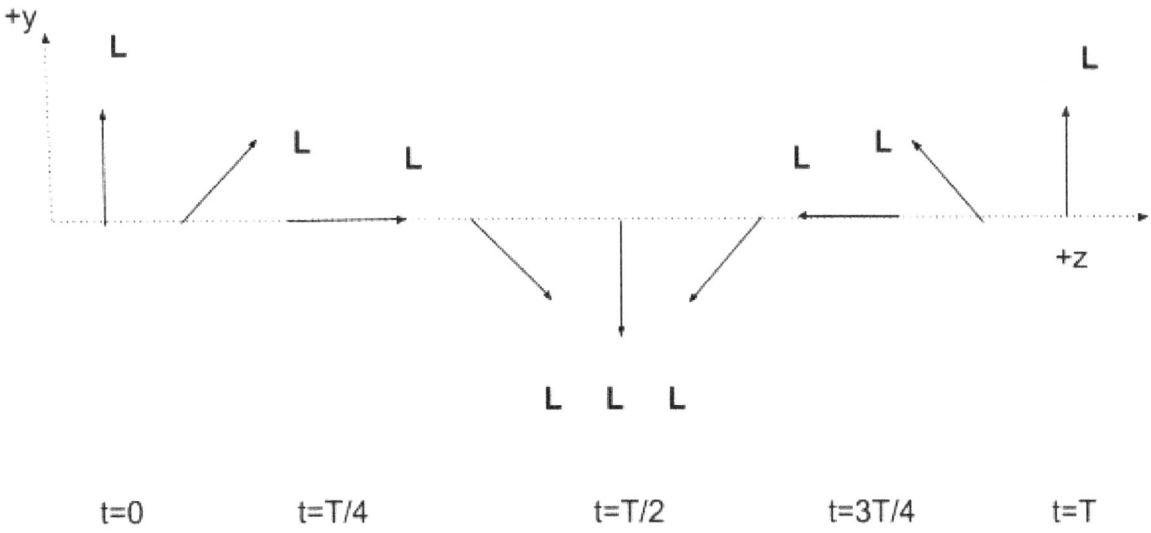

Figure 1: The photon rolls, or 'gyres', along the direction of travel projecting the spin angular momentum vector sinusoidally along the direction of propagation while maintaining a constant polarization and angular momentum; L = h. We could also say that the photon '**spirals**'. At t=0, the plane of the photon spin is *coplanar* with the z-axis.

The photon is massive, yet *inertialess*, because it only has one component of angular momentum, and the projection is along, or parallel to, the direction of propagation.

The force on or due to a photon is

$$F = dp/dt = (h/c) \, dv/dt \qquad (17)$$

The impulse transferred to or from the photon is

$$P = \int_t (h/c)(dv/dt) \, dt = (h/c)(v - v') \qquad (18)$$

We may also construct a minimization principle for the photon;

$$\delta \int W \, dt = \delta \int h \, (dv/dt) \, dt = \delta \int h^{bar}(d\omega/dt) \, dt = 0 \qquad (19)$$

Just like our new 'classical' minimization principle of equation (15), equation (19) involves only angular variables, and the integration is explicit only for the time of the interaction.

Of, course, to apply these equations to physical situations, one must know the forces in order to minimize the work.

The photon has an antiparticle. We will call it the d'oh-ton!
Photons spin to the left and anti-photons spin to the right.

When a photon and a d'oh-ton collide, they form a virtual photon. The photon has a gravitational charge equal to its mass. In our model photons may interact with one another as well as with the leptons.

We can see from Figure 1, that the amount of angular momentum *available for interaction* (and/or transfer), as well as the *ability of the photon to interact at all*, varies sinusoidally in time as the photon travels through space, hence the appearance and behavior of electromagnetic waves.

We will interpret the scheme portrayed in Figure 1 *literally*, and proclaim the photon is a two-dimensional, planar, flat, circular, spinning disk of fundamental angular momentum, h.

The leptons would then be three-dimensional spinning *blobs* of fundamental angular momentum, with three vector components, only one of which may be projected along the direction of travel; hence the leptons acquire *inertia*.

The neutrino:

We imagine the neutrino to be one quantum of action per unit volume of space, resulting in an available, or 'functional', angular momentum of $h^{bar}/2$.

The neutrino is a spinning point particle with three angular momentum vectors.

Only one component of the angular momentum may be projected along the direction of travel, and the total angular momentum 'gimbles' about this direction, as do the other two components.

The gimbling of the neutrino is illustrated in Figure 2.

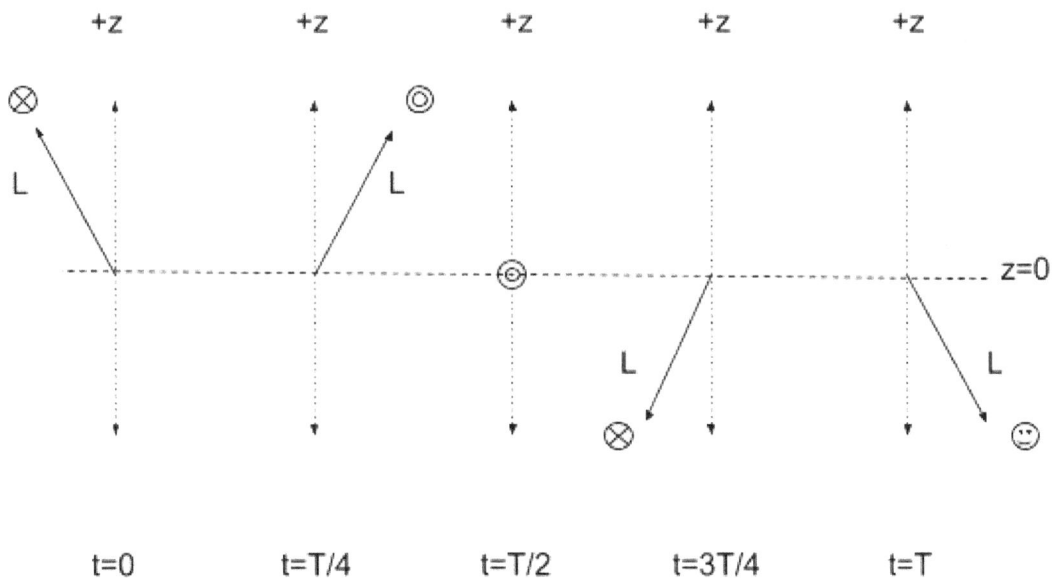

Figure 2: At rest with a lepton traveling the the z-direction. The spin angular momentum vector 'precesses' about the direction of motion, tracing out a closed, three dimensional figure eight. The x symbol represents motion into the page. The dot symbol represents motion out of the page. At time T/2, we see the the angular momentum is *perpendicular* to the direction of travel. (This is when the lepton engages in 'virtual' interactions.)

To represent an antilepton, simply swap the x symbols and the dot symbols.

From Figure 2, we can see that leptons *always* spin to the left, regardless of whether the spin is pointing up or down. That is the power of gimbling!

The neutrino has a gravitational magnetic moment given by (14);

$$\mu = (h^{bar}/2)(1 + \tfrac{1}{2} v^2/c^2 + \tfrac{3}{8} v^4 c^4 + ...) \qquad (20)$$

The magnetic moment is dependent on the particle velocity.

The neutrino magnetic moment may be considered to be a functional 'negative gravitational charge'.

Two neutrinos will attract one another due their mass or gravitational charge. However, they cannot collide, due to the repulsion arising from their identical spins

Identical particles repel one another, and cannot collide, because they spin in the same direction. Particle antiparticle pairs attract one another, because they *can* collide, merging to form new particles. ;)

The electron:

In our model, the existence of the electron is due to the fundamental and irreducible phenomenon of electromagnetic induction (14). In this model, spinning mass gives rise to the electric charge and the electron magnetic moment.

The mass of the electron is the mass of the neutrino multiplied by the *magnitude* of the electric charge;

$$m_e = e\, m_\upsilon \qquad (21)$$

The Coulomb is a 'funny' unit, since it does not involve M, L, or T ! We should like to get rid of it eventually, no offence to Charles de - !, but units of mass, or even no units, would be a better choice. We would then speak of electrical current as mass per second, or even just 'per second' (Hertz)!

The magnetic moment of the electron is velocity dependent like that of the neutrino and is given by;

$$\mu = (e h^{bar}/2m_e)(1 + \tfrac{1}{2} v^2/c^2 + \tfrac{3}{8} v^4 c^4 + ...) \qquad (22)$$

which is just the neutrino magnetic moment multiplied by e/m_e.

The gimbling of the electron is described by the usual wave equation variables;

$$p = h/\lambda \quad ; \quad E = \upsilon h \qquad (23)$$

Now, $\lambda = v/\upsilon$, so

$$p = h\upsilon/v \qquad (24)$$

Then

$$\frac{dp}{dt} = \frac{\partial p}{\partial v}\frac{dv}{dt} + \frac{\partial p}{\partial \upsilon}\frac{d\upsilon}{dt} \qquad (25)$$

$$\frac{dp}{dt} = -\frac{h\upsilon}{v^2}\frac{dv}{dt} + \frac{h}{v}\frac{d\upsilon}{dt} \qquad (26)$$

We also know

$$E = h\upsilon = mc^2 \qquad (27)$$

and,

$$dm/dt = (h/c^2)\, d\upsilon/dt \qquad (28)$$

so equation (26) becomes

$$dp/dt = -(E/v^2)dv/dt + (c^2/v)dm/dt \qquad (29)$$

We *also* know

$$p = mv \qquad (30)$$

and

$$dp/dt = v\, dm/dt + m\, dv/vt \qquad (31)$$

Equating equations (29) and (31) we have

$$v\, dm/dt + m\, dv/vt = -(E/v^2)dv/dt + (c^2/v)dm/dt \qquad (32)$$

We do a boatload of algebra and obtain

$$dm/m = (1 + v^2/c^2)/(1 - v^2/c^2)\, dv/v \qquad (33)$$

Then we realize we are unable to do the integral ...   :(

A homework assignment !

(That's all you get for ten dollars!)   :)

    Math is hard.

                    It takes a universe ...

            Duffman says a lot of things, oh yeah!

                          -- Duffman
                            The Simpsons

                                                    :)

The fundamental lepton:

We have imagined the electron, muon, and tau, to be three manifestations of one fundamental lepton; the tetrahedron. In our model, the tetrahedron has three principal moments of rotational inertia $I_{e,mu,tau}$, corresponding to the three lepton masses.

The gimbling of a particle's moment of inertia, I, about the direction of propagation (in concert with the angular momentum vector, **L** = $\sqrt{3}\,\hbar/2$, *is* relativistic mass.

Curiously, this moment of inertia seems to *not* contribute to particle angular momentum, but *only* to the mass. Let's explore! What could this moment of inertia be?

Consider a non-relativistic particle with velocity v, and kinetic energy

$$T = \tfrac{1}{2} m v^2 \qquad (34)$$

$$\Rightarrow$$

$$T = \tfrac{1}{2} m \lambda^2 \upsilon^2 \qquad (35)$$

Now, if we define the moment of inertia I to be

$$I = m \lambda^2 / (2\pi)^2 \qquad (36)$$

then we can write

$$T = \tfrac{1}{2} I \omega^2 \qquad (37)$$

We define the 'radius' of a point particle to be equivalent to the wavelength. This radius travels 2π radians every wave cycle. The moment of inertia is dependent on the particle wavelength, $I = I(\lambda)$.

Is there a corresponding angular momentum, $L = I\omega$? It seems there should be!

In our model, the magnetic moment increases with velocity and this must be the source. So, we define the angular momentum of an elementary particle to be

$$L = \sqrt{3}\,\hbar/2 + I\omega \qquad (38)$$

and

$$E = m_0 c^2 + \tfrac{1}{2} I \omega^2 \qquad (39)$$

We can also solve for the 'rest frequency' of an electron;

$$m_e c^2 = \hbar \omega_0 \tag{40}$$

$$\omega_0 = m_e c^2 / \hbar = (2\pi c)/(h/m_e c) \tag{41}$$

$$\omega_0 = 2\pi c / \lambda_e \tag{42}$$

where λ_e is, of course, the Compton wavelength of the electron, which we also take to be the 'rest radius'. The uncertainty principle can provide us with the 'rest momentum',

$$\lambda_e \Delta p_e \approx \hbar \tag{43}$$

We can understand these uncertainties, since, even an electron 'at rest', is gimbling, with the radius varying sinusoidally, and the momentum apparently flipping direction at the same rate.

In a similar manner, we can assign a functional radius to the photon as well.

Now, we can offer a **bold**, new hypothesis concerning the double slit experiment.

Perhaps, somehow, when the radius of a particle is on the order of the slit separation, the particle will pass through one slit, whilst 'clipping' or 'grazing' the other slit, altering its angular momentum. Somehow ...

But, back to the tetrahedron, which may be merely metaphorical !

We can use the SU(3) rotation matrices of the standard model to operate on our inertia tensor, and translate from one lepton flavor to another. We will refer to these as rotations in *charge isospin* space.

In addition, we imagine that rotations in 'weak isospin' or SU(2) space, transform the charged lepton into its corresponding neutrino.

Since our model does not involve the weak charge, we must reluctantly rename the concept of weak isospin. We shall now call weak isospin, *mass isospin*.

These ideas are summarized in Table 1.

The leptonic table:

LEPTONS   ANTI-LEPTONS

electron	electron neutrino	PARITY ⇔	electron antineutrino	positron
⇐	CHARGE	MASS ⇕	CHARGE	⇒
muon	muon neutrino	PARITY ⇔	muon antineutrino	anti-muon
⇐	CHARGE	MASS ⇕	CHARGE	⇒
tau	tau neutrino	PARITY ⇔	tau antineutrino	anti-tau
⇔	mass isospin	charge isospin ⇕	mass isospin	⇔

TABLE 1: The leptons and their interrelations; or the kleptogenesis of the leptoquarks.

Any lepton can be 'generated' from any other by the appropriate applications of the parity operator, the mass isospin operator, and our newly proposed 'charge isospin' operator.

The electron, muon, and tau have the same electric charge, but different masses.

Each lepton and its neutrino has the 'same mass', except for an additional (multiplicative) electric charge.

The parity operator transforms particles into antiparticles.

That is the premise of the universal model.

That is the power of symmetry !   :)

Now, let's give the photon a moment of inertia and a magnetic moment. We can see from equations (20 and (22) that a particle's magnetic moment is the angular momentum times a multiplicative constant. The photon angular momentum is h, so the magnetic moment is

$$\mathbf{m}_\gamma = h\,\mathbf{i} \quad ; \quad \mathbf{i} = \mathbf{v}/|\mathbf{v}| \quad ; \quad v = c \tag{44}$$

The mass of the photon is given by equation (16); we insert this into equation (36) to obtain the moment of inertia of the photon

$$I_\gamma = (h\nu/c^2)(\lambda^2/4\pi^2) \tag{45}$$

Symmetry

The electromagnetic charge:

Interactions conserve spin, mass, and charge. In this section, we will generalize the concepts of electric charge and mass, to include all conserved quantities.

Therefore, we define the electromagnetic charge to be

$$Q_{EM} = e \hbar / 2c \, \mathbf{s} \qquad (46)$$

and the mass charge to be

$$Q_{MASS} = m \hbar / 2c \qquad (47)$$

These quantities must be conserved at every 'vertex'.

We are still musing over the potential utility of these definitions …

They are summarized in Table 2.

Force	Coupling constant	Conserved current	Rotation basis	Conserved charge
Electricity	e/m_e	Mass isospin	e, μ, τ	$e*\hbar/2*c\,\mathbf{s}$
Gravity	m_e/e	Charge isospin	e, ν_e	$m*\hbar/2*c$

TABLE 2: Table of coupling constants, conserved currents, and charges: Note: $m_e/e = m_\nu$

Muon decay::

In our model, muon decay *involves no photons* (or any other 'gauge' boson) and is a fundamentally different type of decay from beta decay, or other decays involving a *bound system of particles*. The universal model of muon decay is shown in Figure 3.

There is no 'coupling charge' *per se*, at the vertices, except for mass, for the calculation of lifetimes, decay rates, and probabilities of decay. The only requirement is the 'propagator' be massless. This fixes the phase space for the energy and momentum of the three final state particles, given a particular muon energy.

We imagine the tetrahedron is rotating, or dropping, from the muon axis to the electron axis!

In our model, the muon 'sheds' energy and spin in the form of its neutrino. The energy and spin shed ensures that the ensuing virtual lepton propagator has spin = 0, and is *massless*. In our model, all 'propagators' are massless.

This requirement places constraints on the energy and momentum of the initial and final states in muon decays, and should help to explain the energy hierarchy of the three particle families.

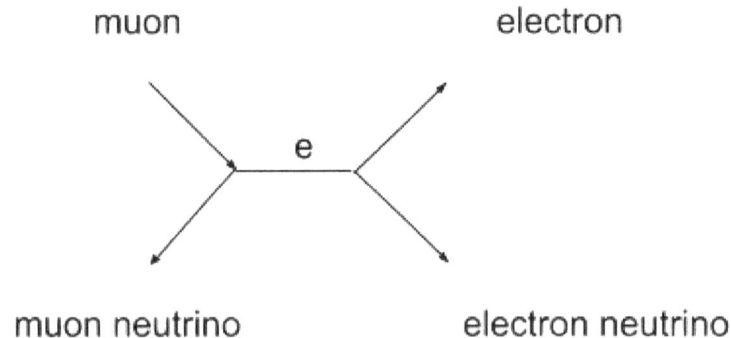

FIGURE 3: Muon decay. The 'propagator' is a generic, 'virtual' lepton, e/mu/tau.

Pilot wave theory:

Our model of electron propagation may be considered a "reverse pilot wave theory". Let us call it the 'pilot particle theory', or perhaps, the 'pilot velocity vector theory'.

In our model, the electron travels a well defined path dictated by the direction of the velocity and/or momentum vector. The velocity *and* momentum of the electron are *completely and solely determined* by the energy, or *frequency*, of the particle.

The electron angular momentum vector gimbles about the 'center of momentum', varying sinusoidally in the ability to engage in 'real' and/or 'virtual' interactions.

Every time the "electron wave function passes through zero", the electron gets updates from all the matter it is interacting with 'at a distance' and it responds accordingly.

This model of electron propagation explains how electrons can 'tunnel' through 'potential barriers". In addition, we can see how an electron with the proper frequency and *phase*, would travel freely through a regular crystalline lattice,

The hydrogen atom:

Let's consider the formation of an hydrogen atom under idealized conditions, as shown in Figure 4. In a two particle universe, we find an electron and a proton *very* far apart. Still, they note a universal attraction; a force that varies with separation, and so they begin moving slowly, inexorably, toward one another, accelerating as they go, gaining kinetic energy due to the mutual forces of one on the other. These forces are equal and opposite, *however* the particle masses are not, so the particles will travel different distances over the allotted interaction time, gaining equal kinetic energy and opposite momenta.

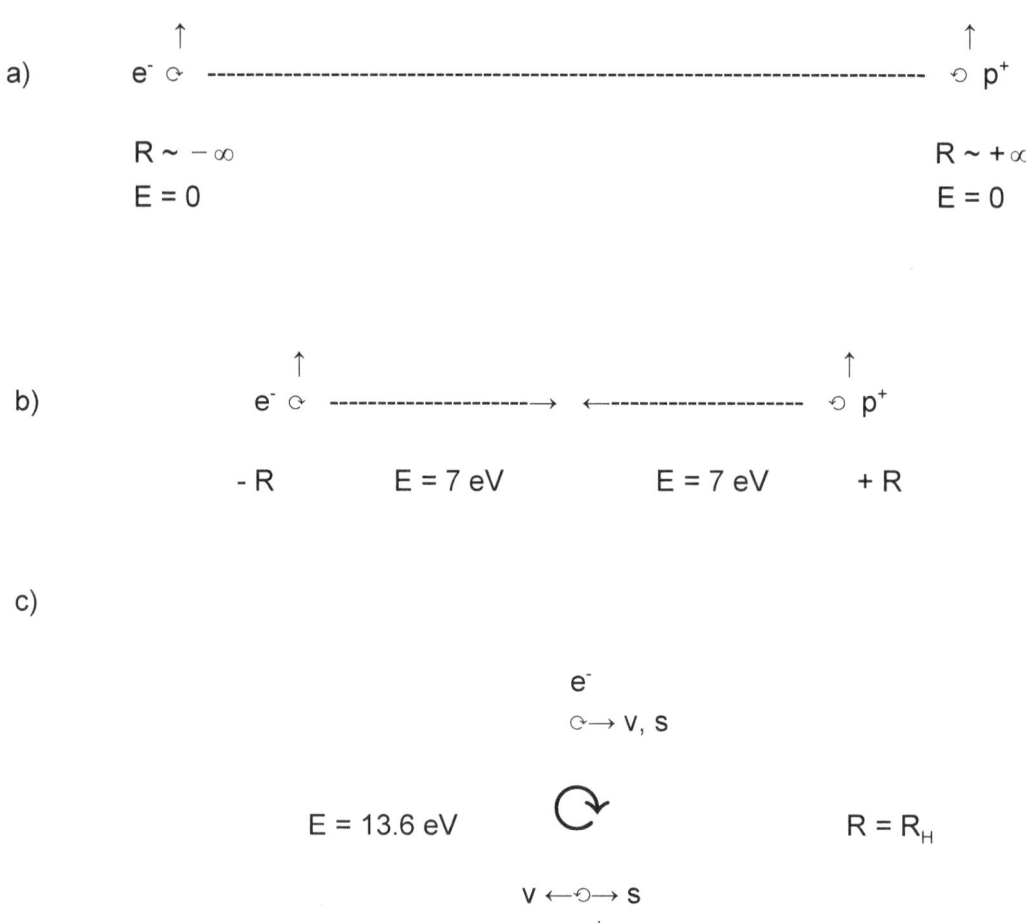

Figure 4: The formation of an hydrogen atom. In a) and b) the spin is shown as pointing up, and the spin *projection* is *not* displayed. In c) we show the projection of the spin and velocity vectors of the particles along, or tangential to, the direction of motion. The electron is matter and the proton is antimatter, so both spins are pointing 'up'!

Given a fortuitous choice of R ~= ∞, we find the electron and proton will have the appropriate energies to form hydrogen when they meet, entering into a mutual orbit about a common center of mass.

The work done on the particles is equal and opposite, resulting in equal and opposite momenta, but a net *gain* in kinetic energy. In the bound system, these kinetic energies correspond to orbital velocities of opposite direction, and they sum to the energy of the hydrogen atom.

We also note the electron and positron have equal and opposite helicities, and these cancel out.

Not only do the 'static' charges attract, but the spins (magnetic moments) of the electron and positron attract too! The two particles cannot be at the same place at the same time, nor can they annihilate, so they orbit.

Perhaps we might think of the vector summation of the work as always summing to zero, and take the scalar sum of the work to represent the total energy of the system.

Now, if we fix our coordinate system at the center of the the proton, the electron orbit will be an ellipse with one focus fixed at the proton center.

In a two particle universe, the hydrogen atom will be *planar*. The electron executes closed elliptical orbits. There are no external forces to change the angular momentum of the particles.

The electron is not a 3D cloud of probability. The proton would be a similar cloud, if so, eh?

Conclusion:

    Potential energy was one the first abstract concepts developed by physicists, and one of the first to assume the mantle of reality; truly believed to be a fundamental aspect of the world; a real, absolutely conserved quantity of energy, somehow taken up and lost by objects during interaction.

    It seems fitting this idea should be the last to fall; this ghostly vestige of *classical* physics.

    Just like our old friend the phlogiston, potential energy turns out to be mass in motion.

    Everything is mass in motion.

    All is action and reaction.

⇦ ⇨ ⇦ ⇨

    Motion is the mode of existence of matter. Never anywhere has there been matter without motion, nor can there be. Motion in cosmic space, mechanical motion of smaller masses on the various celestial bodies, the motion of molecules as heat or as electrical or magnetic currents, chemical combination or disintegration, organic life - at each given moment each individual atom of matter in the world is in one or other of these forms of motion, or in several forms of them at once. All rest, all equilibrium, is only relative, and only has meaning in relation to one or other definite form of motion . . .

Motion without matter is just as unthinkable as matter without motion.

Motion is therefore as uncreatable and indestructible as matter itself;

-- Friedrich Engels
1878

Stimmt!

Everything is connected:

                      Everything is connected !   :)

Every particle in the universe is constantly communicating the following 'information' to each and every other particle, "my energy and momentum with respect to yours is X."

More specifically, each *particle pair* is sharing such *relative* information.

This communication is neither magical or mystical. There is certainly *no room* in these channels for any more complicated message than relative energy, momentum, and *position*.

These messages are being sent over "dedicated lines"; i.e the virtual photon. There are no intervening fields, and the dedicated lines do not split into particle antiparticle pairs "along the way".

The 'distance' between a particle pair is given by the frequency of the virtual photon; a direct consequence of the $1/R^2$ law. The frequency of the virtual photon determines the *rate* of energy and momentum exchange.

                The world is everything that is the case. -- Ludwig Wittgenstein

On math and physics:

    Mathematics does not describe the world, mathematics describes our observations of the world.

    Nature knows neither numbers nor nullity; is as indifferent to the integer as the infinite, noting nothing but matter in motion.

First Principles:

Descending to a more concrete view [of the law of co-operation], we saw that the law sought must be the law of the continuous re-distribution of Matter and Motion. The changes everywhere going on, from those which are slowly altering the structure of our galaxy down to those which constitute a chemical decomposition, are changes in the relative positions of component parts; and everywhere necessarily imply that along with an new arrangement of Matter there has arisen a new arrangement of Motion.

Hence it follows that there must be a law of the concomitant re-distribution of Matter and Motion which holds of every change, and which, by thus unifying all changes, must be the basis of our Philosophy.

-- Herbert Spencer
1862

. . . perhaps there is nothing about so-called educated people and believers in "modern ideas" that is as nauseous as their lack of modesty and the comfortable ignorance of their eyes and hands with which they touch, lick, and finger everything; and it is possible that even among the common people, among the less educated, especially among peasants, one finds today more *relative* nobility of taste and tactful reverence than among the newspaper-reading *demi-monde* of the spirit, the educated.

-- Friedrich Nietzsche
Beyond Good and Evil

Alles klar!

Observations on the quantum mechanical theory of everything:

1. Everything that is not mandatory, is *expressly* forbidden.

2. Modern physics is neither modern *nor* physics, in the *classical* sense;
   i.e. a rational investigation into the nature of matter and interaction based on reason.

3. Just because the wave function is mysterious *to us* (its creators, constructors, and inventors!), does *not* mean it represents some mysterious process, it only means we have yet to divine the process it does represent.

4. The fabric of spacetime was created out of whole cloth;
   the quantum vacuum from thin air.

5. Today's "modern physics" is an atavism; the four forces for the four elements (always four!); spirit guides for motive force.

6. Physics is supposed to be an empirical science, *the* empirical science, yet empiricism lay dying in the ditch (with the cart!), if it is not dead already.

7. The modern physicist is a *sadly* deluded idealist. Disillusioned with the apparent world, which they cannot understand, they invent fantastic realms; many worlds, multiple universes, and manifold dimensions --- these modern mythologists and cynical deists!

8. Interactions are deterministic, but this does not mean they are *predetermined*.

9. After 100 years, modern physics is *still* inchoate, incoherent, and growing inexplicably ever more complicated and confused, a sprawling morras of mysticism and metaphysics, blithely bundled by strapping tape and baling wire, a monstrous Rube Goldberg machine of some cosmic (or perhaps, human?) mind.

    10 What's that fudgy smell? --- renormalization.

    :(

Resources:

Atomic and Quantum Physics
H. Haken, H.C. Wolf

Modern Elementary Particle Physics
Gordon Kane

Classical Dynamics of Particles and Systems
Jerry B. Marion

Foundations of Electromagnetic Theory
John R. Reitz, Frederick J. Milford, Robert W. Christy

Quantum Physics
Rolf G. Winter

Gauge Theories in Particle Physics
I. J. R. Aitchison and A. J. G. Hey

Quarks and Leptons: An Introductory Course in Modern Particle Physics
Francis Halzen, Alan D. Martin

Quantum Field Theory
F. Mandl, G. Shaw

Theoretical Mechanics of Particles and Continua
Alexander L. Fetter, John Dirk Walecka

and

Elementary Modern Physics   (Best Book Ever!)
Richard T. Weidner, Robert L. Sells

Books by Greg Feild: The SInister Universe Series

the pentateuch

1. "A quantum mechanical theory of gravitational interactions"
   CreateSpace Independent Publishing, 8/29/2016

2. "Observations on the quantum mechanical nature of gravity"
   CreateSpace Independent Publishing, 10/8/2016

3. "On gravitation and electric charge"
   CreateSpace Independent Publishing, 10/29/2016

4. "On spin, mass, and charge"
   CreateSpace Independent Publishing, 11/29/2016

5. "On angular momentum, acceleration, and absolute motion"
   CreateSpace Independent Publishing, 1/1/2017

the exegeses

6. "The Sinister Universe"
   CreateSpace Independent Publishing, 3/1/2017

7. "On Parity and Isospin"
   CreateSpace Independent Publishing, 4/11/2017

8. "Reflections on the Sinister Universe"
   CreateSpace Independent Publishing, 5/12/2017

the hermeneutics

9. "On Current Physics"
   CreateSpace Independent Publishing, 6/11/2017

10. "A Critical Examination of Classical and Quantum Mechanical Waves"
    CreateSpace Independent Publishing, 6/18/2017

the gospels   :)

11. "On wave particle duality and the quantum of action"
    CreateSpace Independent Publishing, 7/6/2017

12. "On matter, mass, and motion"
    CreateSpace Independent Publishing, 9/14/2017

13. "On action and reaction"
    CreateSpace Independent Publishing, 9/24/2017

14. "A quantum mechanical theory of everything"
    CreateSpace Independent Publishing, 11/5/2017

the compilations

"The Universal Model of Our Sinister Universe: The First Ten Books"
CreateSpace Independent Publishing, 7/2/2017

"The Canons of the Sinister Universe:
The Last Four Books on the Universal Model of Our World"
CreateSpace Independent Publishing, 11/5/2017

15. "On Interaction
    CreateSpace Independent Publishing, 4/21/2018

About the author:

   I earned a PhD in experimental high energy physics from the Pennsylvania State University working on HERA at DESY in Hamburg, Germany studying photoproduction and deep inelastic scattering in electron-proton collisions.

   I did my postdoctoral studies with Yale University working at Fermilab on the CDF experiment at the Tevatron. My primary research interest was particle hadronization in charmonium production in proton-antiproton collisions.

Life, the Universe and Everything:

        photons; oscillating! tetrahedrons;
        flipping over. orbital decays

        physics!

Appendix A: Newton's laws and the Lorentz force

On action and reaction: From "On Interaction"

The "action" is a well defined quantity and concept in terms of;

1) "Minimizing the action" (i.e. minimizing the time integral of the Lagrangian, T - V, of a system of particles), yielding a quantity with units of (energy) × (time); (E•t).

2) Planck's "quantum of action" with units of E•t, representing one unit (the smallest possible) of 'interaction' that may be exchanged between two particles;

> 'interaction' *defined* as the exchange of (a certain amount of) energy
>   during   (a certain amount of) time.

Everyone knows the units of the action, E•t, are equivalent to the units of angular momentum, kg m^2/s.

Particles interact by exchanging units of action, or units of angular momentum.

Moving angular momentum around takes *work*, and, of course, everyone wants to minimize their amount of work!   "Work smarter, not harder!"

Nature does not like to do work. Nature's motto is: Do as little work as possible !

This is the true anthropic principle.   :)

Newton's third law is often colloquially expressed as "for every action there is an equal and opposite reaction".

Of course, Newton's third law is actually a statement about forces, $\mathbf{F}_2 = -\mathbf{F}_1$, and the term "action" is not well defined. Action seems to be used metaphorically to say that everything that happens to 'object one' during an interaction also happens to 'object two' in a compensating manner, ensuring the conservation of energy, momentum, angular momentum, etc., during the interaction.

This concept of action and reaction is a more general and more inclusive description of the interaction between two bodies, compared to demanding a simple balancing of the central forces, **$F_2 = -F_1$**

In the universal model, Newton's third law will be broadened to include forces as well as "actions".

Just as we 'generalized' the classical Lorentz force (subsuming Newton's universal law of gravitation in the process!), we now 'universalize' Newton's three laws of motion.

1) A body will not experience a change in **momentum**, unless acted on by a force, **F**, generated by a second body (or several other bodies).

2) $\mathbf{F} = d\mathbf{p}/dt = d(m\mathbf{v})/dt = m\, d\mathbf{v}/dt + \mathbf{v}\, dm/dt$ \hfill (A1)

3) The two bodies experience equal and opposite changes in their respective *actions* during the *interaction* (which we have defined as the *exchange of units of action*.)

$$F_1 \times d_1 \times t = -F_2 \times d_2 \times t \quad (A2)$$

and equal and opposite changes in angular momentum due to magnetic forces

$$\Delta L_1 = -\Delta L_2 \quad (A3)$$

We now have equivalent and compatible definitions for "the action" in both the classical and quantum formulations of our universal model.

Newton's universal law of gravitation is *replaced* by the more general, and *properly normalized*, equation (A9), and all calculations are done in the universal inertial reference frame of Figure A1.

The concept of 'minimizing the action' is already the mathematical bridge between classical and quantum physics, and remains so in the universal model, although we expect all Lagrangians to eventually be formulated strictly in terms of ΔT !

Otherwise, we like to avoid the use of the concept of potential energy for formal conceptions and definitions, and save it for the the practical applications of calculating the energy states of real world systems, where various approximations are necessary.

So, in our model, we envision the following substitutions; for the total energy

$$E_{TOT} = T + V = \text{const.} \quad \Rightarrow \quad m = T + m_{rest} \tag{A4}$$

and for the Lagrangian

$$L = T - V \quad \Rightarrow \quad L = T - m_{rest} \tag{A5}$$

where T includes *all* kinetic energy; linear, rotational, etc.

Mass *is* resistance to acceleration. If a mass is successfully accelerated, it tries to shed the extra energy, by emitting photons. Mass does not like to be accelerated!

Nature minimizes the action during an interaction, by keeping the 'relativistic' masses of the particles involved, as small as possible.

Nature achieves change (as it strives for an ever elusive stability) by doing the least amount of work possible, against all possible constraints, and within the boundary conditions fixing the closed system and defining the nature of the interaction.

To do the least amount of work, nature must keep acceleration to a minimum.

For any closed system, the sum of the individual energies divided by the total energy of the system, is equal to one; and we may write, symbolically;

$$(T + m_{rest}) / m = 1 \tag{A6}$$

or more concretely,

$$((\Sigma_i T^i) + (\Sigma_i m^i_{rest})) / (\Sigma_i m^i) = 1 \tag{A7}$$

In this formulation, we see all conservative interactions must satisfy one constraint

$$\Sigma_i (\Delta T_i) = 0 \tag{A8}$$

During any interaction, or evolution of any closed dynamical system, the sum of the changes in the kinetic energy of the particles involved, must add to zero.

In our model, _only relative changes matter_; relative changes in velocity and kinetic energy.

The universal reference frame: From ""On Interaction"

In our model, the force between two bodies orbiting one another, normalized by the total energy of the system, is (14)

$$F/E_{TOT} = K^*(c/R)^2 \mu - K^*(\mu v^2/R^2) - K^*(l^2/\mu R^3) \tag{A9}$$

We may write this symbolically as

$$F_{universal} = F_{central} + F_{coriolis} + F_{centrifugal} \tag{A10}$$

These are all the "force terms". _However,_ there is a contribution to the overall energy of the system, corresponding to the interaction of the spin/angular momentum (σ, l) of one object, with the 'magnetic force vector' of the other object.

$$E_{SPIN}/(E_{TOT}) = (l_1 \cdot B_2)/(m1+m2) + (l_2 \cdot B_1)/(m1+m2) \tag{A11}$$

$$E_{universal} = E_{central} + E_{coriolis} + E_{centrifugal} + E_{spin} \tag{A12}$$

These calculations must be performed in the universal reference frame as shown in Figure A1.

concatenating files is hard!

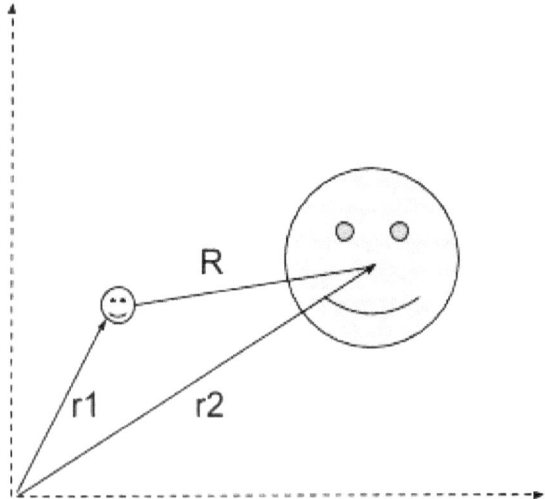

Figure A1: A cartesian coordinate system for the study of planetary motion. The coordinate system is in reference to the fixed background of space. The mutual force between the two bodies is a function of *all* relative velocities between the masses comprising the the two bodies.

The Lorentz force: From "A Quantum Mechanical Theory of Everything"

It is the premise of our model that the motion of mass is the basis of all interaction and all physical phenomena. This assumption was necessary in order to afford the newly massive neutrino a magnetic moment.

For this reason, we find the charge to mass ratio, e/m_rest, of a *charged* particle to be the proper and fundamental *coupling* constant of electromagnetism (long, a popular idea, amongst many!), rather than the electric charge, e, alone, and the *relativistic* mass of the particle to be the actual *coupling charge*.

So, in our model, the complete relativistic Lorentz force on a particle due to a specified distribution of charge and mass is given by

$$\mathbf{F} = (me')*\mathbf{E} + (me')*\mathbf{v}\times\mathbf{B} - m*\mathbf{F\_g} - m*(\mathbf{v}\times\mathbf{B\_g}) \qquad (A13)$$

where m is the relativistic mass of the particle, e' is the charge to mass ratio of the particle

$$e' = e/m_{rest} \tag{A14}$$

and m_rest is the rest mass of the particle.

The term F_g is the Newtonian force due to gravity and is calculated in the usual way (as tiny as it may be!)

$$F_g = G*m1/R^2 \tag{A15}$$

where R is the distance between our particle and a gravitational source charge of relativistic mass m1.

The term B_g (very tiny!) is the 'gravitational magnetic field' vector, analogous to the electromagnetic term B, and is calculated in a similar way;

$$\mathbf{B_g} = (G/c^2)*m1*(\mathbf{v1xR})/R^3 \tag{A16}$$

where v1 is the velocity of the source particle. The constant giving the strength of the force, G/c^2, is chosen to give our resulting gravitational waves the speed of light.

Equation (A13) for the total force on a particle is manifestly Lorentz invariant due to the common mass factor appearing in all four terms. The electromagnetic terms reduce to the usual ('classical') expression in the low particle velocity limit, where m = m_rest.

The original motivation for 'symmetrizing' the classical Lorentz force with respect to gravity was to prop up a quite separate hypothesis, which was that the neutrino had a magnetic moment, even though the magnetic moment arises solely from the 'rotating' motion of its mass.

However, this symmetrization allows for a classical description of gravitational waves *and* bolsters our conclusion from quite separate arguments (1) concerning a gravitationally bound state of two neutrinos, that the graviton is a massless, spin 1, boson just like the photon.

In fact, we were forced to conclude that the graviton and the photon were the same particle for lack of any distinguishing characteristics! (3).

The gravitational waves would continue the electromagnetic spectrum just where the radio waves start to fade away ...

As we constructed our model, we found (or required, demanded) that the gravitational and electromagnetic interactions have the same structure, mechanism, and behavior; the only difference being the value of the electric charge.

In addition, we supposed (postulated) that the gravitational and electromagnetic interactions have the same structure, mechanism, and behavior for both macroscopic and microscopic systems; *the only difference being quantization.*

In this spirit let's have a closer look at our new Lorentz force! Let's rewrite equation (A13) as

$$\mathbf{F} = m*(e/m\_rest)*\mathbf{E} + m*(e/m\_rest)*(\mathbf{v}\times\mathbf{B}) + m*\mathbf{F}\_g + m*(\mathbf{v}\times\mathbf{B}\_g) \qquad (A17)$$

where we are not going to worry about the relative plus and minus signs between the electromagnetic and gravitational terms for the moment.
Factoring out the constant coupling charge, e/m_rest, we have

$$\mathbf{F} = (e/m\_rest)(m*\mathbf{E} + m*(\mathbf{v}\times\mathbf{B})) + (m*\mathbf{F}\_g + m*(\mathbf{v}\times\mathbf{B}\_g)) \qquad (A18)$$

We see that the electromagnetic and gravitational forces now have exactly the same form and they both depend on and vary with the total relativistic mass energy of a particle.

For the electromagnetic terms, e/m_rest is *the* fundamental coupling *constant.*

The charge to rest mass ratio of a particle is now the fixed fundamental coupling constant, rather than e or alpha, and needn't be involved in calculations except as a multiplicative factor. We note, when $m = m\_rest$, the electromagnetic and gravitational forces are totally 'decoupled'.

Now, let's take a closer look at the new mass dependence of the Lorentz force as described by equation (A18). For simplicity, we will focus only on the two electromagnetic terms. The force on a massive charged particle (e.g. an electron) due to a specified distribution of mass and charge is then

$$\mathbf{F} = (e/m_e)(m*\mathbf{E} + m*(\mathbf{v}\times\mathbf{B})) \qquad (A19)$$

If we consider the interaction between two identical electrons, we must assume one provides the field for the other, and equation (9) becomes

$$\mathbf{F} = (e/m_e)(m_1*\mathbf{E} + m_1*(\mathbf{v}_1\times\mathbf{B})) \qquad (A20)$$

Using the well known equations for **E** and **B** due to the second electron, we have

$$\mathbf{F} = (e/m_e)^2(1/R^2)(m_1 m_2(1/4\pi\varepsilon)\mathbf{r} + \mathbf{p}_1\times\mathbf{p}_2\times\mathbf{r}(\mu/4\pi)) \qquad (A21)$$

Some factoring yields

$$F = (e/m_e)^2(c/R)^2(\mu/4\pi)(m_1 m_2 \mathbf{r} + (1/c^2)(\mathbf{p_1} \times \mathbf{p_2} \times \mathbf{r})) \tag{A22}$$

For non-relativistic interactions, we can solve the relationship

$$E = p^2/2m \tag{A23}$$

for the mass of our two electrons, finally yielding

$$F_{1,2} = (e/m_e)^2(c/R)^2(\mu/4\pi)(p_1^2 p_2^2/4E_1 E_2 + (1/c^2)p_1 p_2 \sin\theta) \tag{A24}$$

assuming our two electrons are moving parallel to one another.

The purpose of this exercise was to demonstrate that the classical Lorentz force between two particles can be formulated solely in terms of their energy and momentum.

The classical coupling strength is now given by our new universal coupling parameter, $(e/m\_e)^2$, which can be carried over unscathed and unmolested into quantum mechanical calculations!

We now reintroduce the gravitational terms into equation (A22) and find the the complete, 'classical', relativistic, Lorentz force between two identical electrons

$$F = (G/c^2 - (e/m_e)^2(\mu/4\pi))(c/R)^2)(m_1 m_2 \mathbf{r} + (1/c^2)(\mathbf{p_1} \times \mathbf{p_2} \times \mathbf{r})) \tag{A25}$$

We note that the factor of $(1/c)^2$ dampens the gravitational interaction relative to the electromagnetic interaction. In addition, both the gravitational and electromagnetic 'magnetic' terms are dampened by a factor of $(1/c)^2$ relative to their respective static forces.

Also, notice that the factor of epsilon_0 is 'gone', and we are left with two coupling constants, G and e/m_rest, and a 'space factor', $\mu/4\pi$. The evolution of the force is totally described by the energy and momentum of the two electrons. The factor $(c/R)^2$ describes the time dependence of the interaction, and the time t is *common* to both electrons.

The particles act on one another. There is no 'source' charge.
This description contains no electric or magnetic fields!
All the energy and momentum of the system is carried by the particles.

Nowadays, we characterize this interaction as two electrons trading energy and momentum by virtual photon exchange. Exchange is the key word. There is no emitter and no receiver; just a continual flux of virtual photons between the two.

Actually, a picture I like better, is of one of a continuous virtual photon, constantly changing 'frequency' or 'mass' as the interaction evolves.

Let's look again at the force between two identical electrons as described by equation (A25).

We note, of course, that $\mathbf{F}_1 = -\mathbf{F}_2$. Newton's second law tells us

$$\mathbf{F}_1 = m_1 \mathbf{a}_1 \tag{A26}$$

If we compare equations (15) and (16) we can see that the acceleration of a particle is *independent of its mass*.

$$\mathbf{a}_1 = \text{Function}(m_2, \mathbf{R}) \tag{A27}$$

This is a general result that we used to assume applied only to the gravitational interaction (and is used as an argument in favor of general relativity).

In addition, the relative strengths of the four terms in equation (A25), or the 'four forces of classical physics', are approximately as follows;

electricity = 1 ; magnetism ~ $1/c^2$; gravity ~ G/c^2 ; magnetic gravity ~ G/c^4

I think Maxwell would approve!; except that we have no more need for his fields.

The two electrons exert equal and opposite forces on one another during the interaction, and the evolution of the force is completely described by the (variable) mass-energy of the two electrons, $m_1(\mathbf{r}_1(t))$, $m_2(\mathbf{r}_2(t))$, where the time, t, is *common* to both electrons.

Of course, without the fields there is no mathematical or physical mechanism to explain the interaction of these two particles 'at a distance'.

We like the idea of 'one virtual photon' (which, of course, is a discovery of the field theory model) constantly coupling the particles; a time varying conduit for energy and momentum exchange. Can we infer or derive the virtual photon without resort to field theory?

We will defer this investigation for now.

Let's look at our 'new' Lorentz force in a little more detail. If we define

$$K == (G/c^2 - (e/m_e)^2(\mu/4\pi)) \qquad (A28)$$

then we can write equation (A25) as

$$\mathbf{F} = K*(c/R)^2(m_1 m_2 \mathbf{r} + (1/c^2)(\mathbf{p_1} \mathbf{x} \mathbf{p_2} \mathbf{x} \mathbf{r})) \qquad (A29)$$

Next, we factor out the particle masses from the momentum term

$$\mathbf{F} = K*(c/R)^2(m_1 m_2 \mathbf{r} + (1/c^2)(m_1 m_2 \mathbf{v_1} \mathbf{x} \mathbf{v_2} \mathbf{x} \mathbf{r})) \qquad (A30)$$

$$\mathbf{F} = K*(c/R)^2(m_1 m_2)(\mathbf{r} + (1/c^2)(\mathbf{v_1} \mathbf{x} \mathbf{v_2} \mathbf{x} \mathbf{r})) \qquad (A31)$$

where $\mathbf{r}$ is the unit vector $\mathbf{R}/R$.

Since our two masses form a closed, conservative system, we can 'normalize' our force by dividing by the total energy of the system; $E_{TOT} = m_1 + m_2$.

$$\mathbf{F}/E_{TOT} = K*(c/R)^2 \mu (\mathbf{r} + (1/c^2)(\mathbf{v_1} \mathbf{x} \mathbf{v_2} \mathbf{x} \mathbf{r})) \qquad (A32)$$

where $\mu(\mathbf{R}, d\mathbf{R}/dt) = m_1 m_2/(m_1 + m_2)$ is the reduced mass of the two body system.

In order to investigate the vector cross product term, we will assume our two masses (no longer necessarily electrons) are equal and orbiting one another.

Then we can write

$$\mathbf{F}/E_{TOT} = K*(c/R)^2 \mu (1 - (v^2/c^2)) \qquad (A33)$$

Seriously! and/or

$$\mathbf{F}/E_{TOT} = K*(c/R)^2 \mu - K*(\mu v^2/R^2) \qquad (A34)$$

We will call the second term in equation (A34), the *coriolis* force.

If we recast our force equation into polar coordinates and allow $m_1 \neq m_2$ (i.e. for the study of planetary motion; Kepler's equations), we will pick up the usual *centrifugal* force term, in addition to our new *coriolis* force term.

$$\mathbf{F}/E_{TOT} = K*(c/R)^2 \mu - K*(\mu v^2/R^2) - K*(l^2/\mu R^3) \qquad (A35)$$

Finally, there will be a force term corresponding to the interaction of the spin/angular momentum (σ, **l**) of one object with the 'magnetic force vector' of the other object; F_{spin}.

(Remember, in our model, the relativistic mass of an object is due to the *total* relativistic motion of the mass; including spin.

So, the total Lorentz force, for cosmology for example, will consist of four terms;

$$F_{universal} \sim F_{central} + F_{coriolis} + F_{centrifugal} + F_{spin} \qquad (A36)$$

In our generalized Lorentz force, the coriolis force is not an artifact of the choice of a particular reference frame, but arises from the absolute relative motion of the two bodies.

In our model, the coriolis force is *real*, because all forces are velocity dependent due to the fact that particle mass is velocity dependent; $m = m(\mathbf{r},\mathbf{v})$.

When we speak of the relativistic mass in our model, this includes *all relative velocities*, and not only linear velocities approaching the speed of light.

Similarly, the centrifugal force is now also a real velocity dependent force and is not due to "space time" disturbances as in the general theory of relativity.

Equation (A35) should be 'easy' to generalize to an N body system.

Units and dimensions:

There is essentially only one *variable* of physics; frequency.

Frequency determines the velocity, and hence the linear momentum, mass, and/or kinetic energy of all particles. *Relative* frequencies determines the strength of all particle interactions. We can also characterize macroscopic interactions solely in terms of angular momentum and frequency change.

The fundamental parameters of nature are: t, h, c !

Planck's constant, h, contains *all* the kinematic/dynamical units; M, L, T.

The world is dimensional and our constants should be dimensional too.

For the three coupling constants we choose; G, m_e, e.

The constant ε_0 can be written in terms of G or vice versa. It seems better to cast G in terms of ε_0 (of the vacuum) since ε varies according to the medium of propagation whereas G does not.

As for the electric charge, we suggest and believe it should have the units of mass. This would be an extra "electrical mass" charge that does not contribute to the gravitational mass charge, or its equivalent, the inertial mass.

There are several benefits to this slightly mind sloshing idea. The first being that the formulas for the neutrino magnetic moment, equation (20), and the electron magnetic moment, equation (22), would have the same units; namely, h; a nice symmetry.

Secondly, this would reduce our model to one parameter with units, h, and one variable with units, the particle velocity, v. In addition, the following four quantities would become unitless parameters;

$$v/c, \quad e/m_e, \quad \mu_0, \quad G/\varepsilon_0$$

The Ampere would take on units of mass/second and the 'electrical' current would be

$$I = (e/m_e) \, dm/dt \tag{48}$$

easily allowing treatment of relativistic electrical currents.